THE QURAN & SCIENCE

NIYAH PRESS

THE QURAN & SCIENCE: GUIDANCE FOR MANKIND
WITH SCIENTIFIC DISCOVERIES AND PROPHECIES

Copyright © 2016 by Ghiasuddin Ahmed Khan. All rights reserved.
Printed in the United States of America. No part of this book may
be used or reproduced in any manner whatsoever without written
permission except in the case of brief quotations embodied in critical articles and reviews.

Niyah Press books may be purchased for educational, business, or
sales promotional use. For information, please write to
info@niyahpress.com or see www.niyahpress.com

To contact the family, email quranicscience1@gmail.com

FIRST EDITION

ISBN: 978-1533616708

THE QURAN & SCIENCE

GUIDANCE FOR MANKIND WITH SCIENTIFIC DISCOVERIES AND PROPHECIES

GHIASUDDIN AHMED KHAN

Niyah Press

Detroit, MI

This project was completed and published by the family of Ghiasuddin Ahmed Khan, for we honor and love him. We ask that Allah give him the blessings of this work for now and forever.

DEDICATION

My Nanaba, Mr. Ghiasuddin Ahmed Khan, was a proud, wise, and intelligent man of faith. It is with great honor and Allah's blessing that we have the opportunity to make his dream of publishing this book a reality. Thank you to all who supported this initiative.

The publishing of this book is dedicated to the loving memory and legacy of Nanaba and his dear wife, my Nanima, Mrs. Zeenath Khan.

Nabila Ikram
Granddaughter

CONTENTS

Preface	11
The Laws	14
Part One: Discoveries	17
One: Time	19
Two: Atmosphere	21
Three: Force	23
Four: Allah's Laws	25
Five: Horizons	27
Six: Atoms	29
Seven: Perfection	31
Eight: Creation	33
Nine: Reflecting Moon	34
Ten: Orbits	36
Eleven: Big Bang & Water	39
Twelve: Two Easts and Wests	41
Thirteen: Seas	42
Fourteen: Reduction of outlying parts	44
Part Two: Prophecies	46
Signs of Doomsday	
Warnings	52
Scientific Evidence of Prophecies	54
Author Biography	57
Tribute	59

THE QURAN & SCIENCE

GUIDANCE FOR MANKIND WITH SCIENTIFIC DISCOVERIES AND PROPHECIES

PREFACE

الله

The Arabic word for the Almighty, The One God is Allah. The greatest attraction Islam offers to its followers is its worship of One Supreme God. The conception of God presented by Islam is that He is The Highest and Most Exalted. The Quran sums it up as follows:

قُلْ هُوَ ٱللَّهُ أَحَدٌ
ٱللَّهُ ٱلصَّمَدُ
لَمْ يَلِدْ وَلَمْ يُولَدْ
وَلَمْ يَكُن لَّهُ كُفُوًا أَحَدٌ

"Say: He is Allah, the One;
Allah, the eternally besought of all;
He begetteth not nor was He begotten;
And there is none comparable unto
Him." (Quran 112:1-4)

Thus, God has no father, no mother, no brother, no sister, no wife, no sons, no daughters, no equals, no partners, no antagonists, and no rivals.

He is the sole creator of all that exists (Quran 14:32). He has created and subjected everything to universal and changeless laws. No one has the power to change these laws, which control and guide all aspects of nature and life (Quran 33:62).

It follows that God is present, has knowledge of and power over everything, everywhere and at all times. He was always present everywhere, He is present now everywhere, and He will be present everywhere…forever. There are no outer limits to His presence, nor are there any enclaves or voids within them. The same applies to His knowledge and power.

God's laws are intended to help mankind climb to higher and higher planes of knowledge, achievement, and fulfillment. Man has been given will-power to either work according to these laws, or he can go against them if he so desires. **But there are limits set by God which man must not transgress.** If he transgresses these limits, he suffers. If he goes too far, he may even be destroyed. But if he realizes his mistake and repents, desists from committing it in the future, and prays to God for forgiveness, he will find God most merciful (Quran 2:160, 16:119).

Thus, a Muslim believes that whatever good or ill, gain or loss, that happens to him is certainly from God and His laws, but ultimately, the result of his own good or evil actions.

WHAT ARE THESE LAWS AND WHERE CAN WE FIND THEM?

We Muslims are blessed by God, as He has sent down his book, ***The Holy Quran,*** on us, which conforms and clarifies the revelation in the Bible and the Torah.

The Quran is a guide to show us right from wrong. It is also a mercy on us, if we choose to follow His commandments revealed in the Quran and through His messenger, Prophet Mohammed ﷺ.

Allah has advised us several times in the Quran to read His book carefully and to learn to follow what it means (Quran 3:190-191, 38:29, 39:55), instead of simply repeating it like a parrot without understanding the meaning. Although there is much benefit in listening to, reading and reciting the Quran without understanding, we should always strive to get closer to Allah by learning the meanings of His words.

The verses in the following section were revealed in the Quran about 1400 years ago. Through modern technological advancements, most have been proven to be not only true, but also to prove Allah's omniscience: that He knows everything at all times – in the past, present, and in the future.

The revelations in the Quran are from the Almighty alone, and are not, nor can they ever be, from anyone's imagination.

Part One:

Discoveries

ONE: TIME

وَيَسْتَعْجِلُونَكَ بِٱلْعَذَابِ وَلَن يُخْلِفَ ٱللَّهُ وَعْدَهُۥ ۚ وَإِنَّ يَوْمًا عِندَ رَبِّكَ كَأَلْفِ سَنَةٍ مِّمَّا تَعُدُّونَ

"And they urge you to hasten the punishment. But Allah will never fail in His promise. And indeed, a day with your Lord is like a thousand years of those which you count." Quran (22:47)

The Earth's circumference is approximately 24,000 miles. One complete spin of Earth on its axis takes roughly 24 hours. Thus, the speed of the Earth's spin is approximately 1000 miles per hour.

The speed of light is approximately 186,000 miles per second. It will take a fraction of a second for the light to go around the Earth once.

If traveling with the speed of light in outer space, considering the Earth's spin, and calculating the Earth's spin in seconds, minutes and hours, it is observed that in one day's travel of daylight, the Earth will spin the equivalent of 1000 years approximately.

It has been seen by many, including prophets, saints and others, that supernatural beings move at lightning speed. They appear and disappear in a flash. What speed is this? It may be even faster than the speed of light!

Two: Atmosphere

$$\text{لَقَدْ خَلَقْنَا ٱلْإِنسَـٰنَ فِى كَبَدٍ}$$

"We verily created human beings in an atmosphere." Quran (90:4)

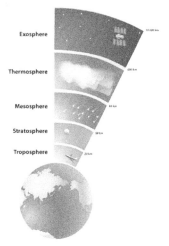

It has been discovered that the Earth has many layers of atmosphere. These serve to protect the globe from the harmful radiation rays of the sun and to save the earth by burning away the great quantities of celestial debris showering upon it with increased speed.

An atmospheric pressure of 14.7 pounds per square inch is maintained, at all times, on every inch of the Earth's surface (like the pressure in an airplane's cabin). Without this pressure there would be no control on stability. Compounded with the high speed of the Earth's rotation, a lack of atmospheric pressure would cause all on Earth to fly away, drifting into outer space, out of control.

THREE: FORCE

يَـٰمَعْشَرَ ٱلْجِنِّ وَٱلْإِنسِ إِنِ ٱسْتَطَعْتُمْ أَن تَنفُذُوا۟ مِنْ أَقْطَارِ ٱلسَّمَـٰوَٰتِ وَٱلْأَرْضِ فَٱنفُذُوا۟ لَا تَنفُذُونَ إِلَّا بِسُلْطَـٰنٍ

"O company of jinn and mankind, if you are able to pass beyond the regions of the heavens and the earth, then pass. You will not pass except by authority [from Allah]."
Quran (55:33)

The Arabic term which here translates to "authority" can also be translated as "a force."

Only in the 20th century has mankind discovered that to travel to outer space, a great force of rocketry is required to lift spaceships from gravity's tremendous grip.

Four: Allah's Laws

فَأَقِمْ وَجْهَكَ لِلدِّينِ حَنِيفًا فِطْرَتَ اللَّهِ
الَّتِي فَطَرَ النَّاسَ عَلَيْهَا لَا تَبْدِيلَ
لِخَلْقِ اللَّهِ ذَلِكَ الدِّينُ الْقَيِّمُ وَلَكِنَّ
أَكْثَرَ النَّاسِ لَا يَعْلَمُونَ

"So set thy purpose (O Muhammad) for religion as a man by nature upright – the nature (framed) of Allah, in which He hath created man. There is no altering (the laws of) Allah's creation. That is the right religion, but most men know not."
Quran (30:30)

This is one of the great subjects, and may take many human lifetimes to cover. But for now, as the laws of Allah become better understood, it is possible to plan and carry out research and projects on land, in the sea, and in outer space for the welfare of living beings.

From the simplest to the most sophisticated computers and equipment, nothing could have been invented in the absence of absolutely firm natural laws. For instance, we know electricity is created by turning an electric generator. If one were to ask why turning the generator produces electricity, the answer would be: the wiring is formed in such a way that it creates an electromagnetic field and electric current. But if one continued to ask 'why' with every answer, the final reply of even the greatest scientist or engineer would be a shrugging of their shoulders saying that it just happens. We know why it happens – because of the firm laws of Allah, which cannot be altered unless Allah permits it.

Five: Horizons

سَنُرِيهِمْ ءَايَٰتِنَا فِى ٱلْءَافَاقِ وَفِىٓ أَنفُسِهِمْ حَتَّىٰ يَتَبَيَّنَ لَهُمْ أَنَّهُ ٱلْحَقُّ ۗ أَوَلَمْ يَكْفِ بِرَبِّكَ أَنَّهُۥ عَلَىٰ كُلِّ شَىْءٍ شَهِيدٌ

"We will show them Our manifest signs (proofs) in the horizons of the universe and within their own selves, until it will become manifest to them that it (the Qur'an) is indeed the truth. Is it not sufficient (as proof) that your Lord is a witness over all things (just as He is witnessed to by all things)?" Quran (41:53)

After the revelation of the Holy Quran over 1400 years ago – a very short period for Allah Ta'ala – we have been graced with the mind development for knowledge and inventions of equipment such as rockets, spaceships, and such telescopes to look deep into the known universe and enjoy the wonders of creation. These discoveries have confirmed that Allah

Ta'ala's creation is limitless by human standards; there is no limit to the number of stars, galaxies and planets in outer space.

We have further been graced with various medical equipment such as computer tomographic (CT) scanners and magnetic resonance imaging (MRI) systems which permit us to marvel at the functions of the greatest creation: human beings. Ultrasound machines help diagnose illnesses by allowing us to look inside all living bodies without the need for surgery. All these developments evolved mostly in the 20th century.

There seems to be no limit to the number of microscopic beings as electronic and atomic microscopes continue to discover smaller and smaller objects. For instance, by use of the electronic microscope, scientists have discovered a germ living on the human eyelash. This germ resembles an elephant with legs and trunk, and although the reason for its existence is yet unknown, it may very well be carrying even smaller organisms, who in turn may carry other creatures; all of these look up to Allah Ta'ala for safety and sustenance.

Six: Atoms

وَمَا تَكُونُ فِى شَأْنٍ وَمَا تَتْلُواْ مِنْهُ مِن قُرْءَانٍ وَلَا تَعْمَلُونَ مِنْ عَمَلٍ إِلَّا كُنَّا عَلَيْكُمْ شُهُودًا إِذْ تُفِيضُونَ فِيهِ وَمَا يَعْزُبُ عَن رَّبِّكَ مِن مِّثْقَالِ ذَرَّةٍ فِى ٱلْأَرْضِ وَلَا فِى ٱلسَّمَآءِ وَلَآ أَصْغَرَ مِن ذَٰلِكَ وَلَآ أَكْبَرَ إِلَّا فِى كِتَٰبٍ مُّبِينٍ

"In whatever activity you (Muhammad) may be engaged, and whichever part of the Quran you recite, and whatever deed you do, We are witness to it when you are engaged in it. And not even an atom's weight in the earth or in the sky escapes your Lord, nor what is less than or greater than that, but it is (written) in a clear Book."
Quran (10:61)

Only recently, after great research, has it come to be known that microscopic atoms can also be broken down into further, smaller particles. Before, the atom was considered the smallest weight, and thought to be unbreakable.

SEVEN: PERFECTION

وَتَرَى ٱلْجِبَالَ تَحْسَبُهَا جَامِدَةً وَهِىَ تَمُرُّ مَرَّ ٱلسَّحَابِ صُنْعَ ٱللَّهِ ٱلَّذِىٓ أَتْقَنَ كُلَّ شَىْءٍ إِنَّهُۥ خَبِيرٌۢ بِمَا تَفْعَلُونَ

"You will see the mountains, which you reckoned were solid, slip away just as clouds slip away, through the handiwork of God Who consummates everything. He is Informed about whatever you are doing."
Quran (27:88)

Allah Ta'ala, who has power and strength over all things, has created every part of the universe proportionately strong, so that it requires great effort from mankind to alter. For instance, it takes the force of a great explosion to detonate an atomic bomb – which in fact breaks down the atom, the microscopic particle that cannot even be seen by the naked eye. This shows the magnitude of the strength of the bonds that keep everything in control and order.

EIGHT: CREATION

ثُمَّ ٱسْتَوَىٰ إِلَى ٱلسَّمَاءِ وَهِيَ دُخَانٌ فَقَالَ لَهَا وَلِلْأَرْضِ ٱئْتِيَا طَوْعًا أَوْ كَرْهًا قَالَتَا أَتَيْنَا طَائِعِينَ

"Then turned He to the heaven when it was smoke, and said unto it and unto the earth: Come both of you, willingly or loth. They said: We come, obedient." Quran (41:11)

With the help of the largest telescope launched into outer space, great masses of gas and dust and other heavenly bodies can be seen floating in the cosmos. Scientists have also concluded with certainty that new stars and galaxies are still being created, as our universe was once created, from dust and smoke.

NINE: REFLECTING MOON

تَبَارَكَ ٱلَّذِى جَعَلَ فِى ٱلسَّمَآءِ بُرُوجًا وَجَعَلَ فِيهَا سِرَاجًا وَقَمَرًا مُّنِيرًا

"Blessed be He Who placed in the heaven mansions of the stars, and placed therein a great lamp and a luminous moon." Quran (25:61)

Atomic reactions and explosions on the sun's surface sending waves of flames and

winds to great distances in space, show that it burns like a lamp, giving out light and energy.

Man's travel to the moon have confirmed the presence of mica spread over its surface. The moon rotates just enough for its mica covered surface to act as a reflector, receiving sunlight and directing it to Earth at all times.

TEN: ORBITS

وَهُوَ ٱلَّذِى خَلَقَ ٱلَّيْلَ وَٱلنَّهَارَ وَٱلشَّمْسَ وَٱلْقَمَرَ ۖ كُلٌّ فِى فَلَكٍ يَسْبَحُونَ

"And He it is Who created the night and the day, and the sun and the moon. They float, each in an orbit."
Quran (21:33)

اللَّهُ الَّذِي رَفَعَ السَّماواتِ بِغَيرِ عَمَدٍ تَرَونَها ۖ ثُمَّ استَوىٰ عَلَى العَرشِ ۖ وَسَخَّرَ الشَّمسَ وَالقَمَرَ ۖ كُلٌّ يَجري لِأَجَلٍ مُسَمًّى ۚ يُدَبِّرُ الأَمرَ يُفَصِّلُ الآياتِ لَعَلَّكُم بِلِقاءِ رَبِّكُم توقِنونَ

"It is Allah who erected the heavens without pillars that you [can] see; then He established Himself above the Throne and made subject the sun and the moon, each running [its course] for a specified term. He arranges [each]

matter; He details the signs that you may, of the meeting with your Lord, be certain." Quran (13:2)

The earth's diameter is approximately 8,000 miles, where the highest mountain peak is approximately 29,000 feet and the deepest trench is approximately 36,000 feet. Using these figures, the calculated ratio of fineness of the Earth's surface comes to 0.0007.

In other words, Allah's creation of Earth (not to mention other mansions of stars as well) come within 0.0007 of an inch to perfect smoothness. This finest finish on any surface cannot be obtained without high tech, computer-aided machines, or in the vacuum of outer space where there is no pull of gravity.

Due to the fine surfaces of these heavenly bodies, and as there is no abrasive resistance to them in outer space, they can float forever by human standards, or until a period set by Allah Ta'ala. Scientists have confirmed the phenomenon that the sun's energy and the earth's inner heat are both gradually reducing, and in the future, they will extinguish and collapse.

ELEVEN: BIG BANG & WATER

أَوَلَمْ يَرَ ٱلَّذِينَ كَفَرُوٓاْ أَنَّ ٱلسَّمَٰوَٰتِ وَٱلْأَرْضَ كَانَتَا رَتْقًا فَفَتَقْنَٰهُمَا ۖ وَجَعَلْنَا مِنَ ٱلْمَآءِ كُلَّ شَىْءٍ حَىٍّ ۖ أَفَلَا يُؤْمِنُونَ

"Have not disbelievers known that the heaven and earth were of one mass, then we parted them, and made every living thing of water. Will they not believe?" Quran (21:30)

Scientists have confirmed that many billions of years ago the universe was created from one point and with an explosion, now known as the "big bang", and that from the big bang, all stars and planets were created. In other words, as Allah Ta'ala says, everything started as one entity that divided.

We also know that the reproduction of living things happens with liquid, and that every fetus remains in water within the mother's womb until its birth. Consequently, exploring living cells at the molecular level has uncovered the fact that 99% of these molecules are water and that it comprises about 70% of the weight of any living organism.

TWELVE: TWO EASTS AND WESTS

رَبُّ ٱلْمَشْرِقَيْنِ وَرَبُّ ٱلْمَغْرِبَيْنِ

"Lord of the two Easts, and lord of the two Wests." Quran (55:17)

We know that East and West are the directions from which the sun rises and sets, respectively. While the Earth spins on its own axis, it also rotates around the sun. This rotation creates the winds and the seasons of the planet. Notice that during the summer and winter seasons, the sides from where the sun rises and sets differ, confirming two rising and two setting places of the sun, two Easts and two Wests.

Thirteen: Seas

وَهُوَ ٱلَّذِى مَرَجَ ٱلْبَحْرَيْنِ هَـٰذَا عَذْبٌ فُرَاتٌ وَهَـٰذَا مِلْحٌ أُجَاجٌ وَجَعَلَ بَيْنَهُمَا بَرْزَخًا وَحِجْرًا مَّحْجُورًا

"It is He Who has let free the two bodies of flowing water: One palatable and sweet, and the other salt and bitter; yet has He made a barrier between them, a partition that is forbidden to be passed."
Quran (25:53)

Chemical analysis has shown that water of different oceans and seas have different characteristics, and though they meet, they do not mix; they maintain their own singularities.

During one of Columbus' voyages to America, his crew, who were desperately thirsty, found a river of soft drinkable water in the middle of the hard waters of the Atlantic Ocean!

Fourteen: Reduction of Outlying Parts

أَوَلَمْ يَرَوْاْ أَنَّا نَأْتِى ٱلْأَرْضَ نَنقُصُهَا مِنْ أَطْرَافِهَا وَٱللَّهُ يَحْكُمُ لَا مُعَقِّبَ لِحُكْمِهِ وَهُوَ سَرِيعُ ٱلْحِسَابِ

"Have they not seen how We deal with the earth, eroding it at its extremities? God decides; no one can modify His decision. He is swift in calling to account." Quran (13:41)

It appears that distances on earth are becoming smaller and time is becoming shorter as the days go by. The reasons for this are twofold:

1) Development of high and ultra-high speed transportation – beginning with the invention of wheels and evolving to sophisticated rockets and spaceships – reduce travel times between great distances to the fraction of what they used to be. It is as though the outer lying parts of the earth are drawing closer and closer

to each other, or the distances between them have gradually reduced.

2) Development of telecommunication has progressed from the telegraph, to the radio to the transistor, telephone, television and now the satellite. Instant, live communication from any distance is now possible. This proves Allah Ta'ala's omniscience when He says that the outer lying parts of Earth are being gradually drawn closer.

Part Two:

Prophecies

The last day on earth, also known as "doomsday," is approaching fast according to the Holy Quran. It narrates that when the earth will be in turmoil of distrust, cruelty, usury, adultery, wars of greed, and filled with the killing of innocent people, Allah Ta'ala will send warnings against these disorders. These warnings include chemically induced diseases, earthquakes, tsunamis, and climate change.

SIGNS OF DOOMSDAY

ظَهَرَ ٱلْفَسَادُ فِى ٱلْبَرِّ وَٱلْبَحْرِ بِمَا كَسَبَتْ أَيْدِى ٱلنَّاسِ لِيُذِيقَهُم بَعْضَ ٱلَّذِى عَمِلُواْ لَعَلَّهُمْ يَرْجِعُونَ

"Corruption does appear on land and sea because of the evil which men's hands have done, that He may make them taste a part of that which they have done, in order that they may return." Quran (30:41)

—

فَهَلْ يَنظُرُونَ إِلَّا ٱلسَّاعَةَ أَن تَأْتِيَهُم بَغْتَةً ۖ فَقَدْ جَاءَ أَشْرَاطُهَا ۚ فَأَنَّىٰ لَهُمْ إِذَا جَاءَتْهُمْ ذِكْرَىٰهُمْ

"Await they aught save the Hour, that it should come upon them unawares? And the beginnings thereof have already come. But how, when it hath come upon them, can they take their warning?" Quran (47:18)

—

وَمَا يَنظُرُ هَٰؤُلَاءِ إِلَّا صَيْحَةً وَاحِدَةً مَّا لَهَا مِن فَوَاقٍ

"They wait for but one shout, there will be no stoppage once started." Quran (38:15)

—

وَيَوْمَ يُنفَخُ فِى ٱلصُّورِ فَفَزِعَ مَن فِى ٱلسَّمَٰوَٰتِ وَمَن فِى ٱلْأَرْضِ إِلَّا مَن شَاءَ ٱللَّهُ ۚ وَكُلٌّ أَتَوْهُ دَاخِرِينَ

THE QURAN AND SCIENCE

(٨٧) وَتَرَى ٱلْجِبَالَ تَحْسَبُهَا جَامِدَةً وَهِيَ تَمُرُّ مَرَّ ٱلسَّحَابِ صُنْعَ ٱللَّهِ ٱلَّذِى أَتْقَنَ كُلَّ شَىْءٍ إِنَّهُ خَبِيرٌۢ بِمَا تَفْعَلُونَ

"And (remind them of) the Day when the Trumpet will be blown, and all who are in the heavens and the earth will be afraid, except those whom Allah wills. And all shall come unto Him, humbled (87). You will see the mountains, which you reckoned were solid, slip away just as clouds slip away, through the handiwork of God Who consummates everything. He is Informed about whatever you are doing." Quran (27:87, 88)

—

يَوْمَ تَرْجُفُ ٱلرَّاجِفَةُ (٦) تَتْبَعُهَا ٱلرَّادِفَةُ

"On the day when the first trumpet resounds and the second follows it." Quran (79:6, 7)

$$\text{يَوْمَ تَرْجُفُ ٱلْأَرْضُ وَٱلْجِبَالُ وَكَانَتِ ٱلْجِبَالُ كَثِيبًا مَّهِيلًا}$$

"On the day the earth and mountains will rumble, and the mountains shall spill over as if they had been turned into sand." Quran (73:14)

—

$$\text{فَٱرْتَقِبْ يَوْمَ تَأْتِى ٱلسَّمَآءُ بِدُخَانٍ مُّبِينٍ (١٠) يَغْشَى ٱلنَّاسَ ۖ هَٰذَا عَذَابٌ أَلِيمٌ}$$

"Therefore watch (O Muhammad) for the day when the sky will produce visible smoke (10). That will envelop the people. This will be a painful torment." Quran (44:10, 11)

—

فَإِنَّمَا هِىَ زَجْرَةٌ وَأحِدَةٌ (١٣) فَإِذَا هُم بِٱلسَّاهِرَةِ

"Surely there will be but one shout (13). And lo! they will be awakened."
Quran (79:13, 14)

WARNINGS

إِنَّ فِى خَلْقِ ٱلسَّمَٰوَٰتِ وَٱلْأَرْضِ وَٱخْتِلَٰفِ ٱلَّيْلِ وَٱلنَّهَارِ لَءَايَٰتٍ لِّأُو۟لِى ٱلْأَلْبَٰبِ (١٩٠) ٱلَّذِينَ يَذْكُرُونَ ٱللَّهَ قِيَٰمًا وَقُعُودًا وَعَلَىٰ جُنُوبِهِمْ وَيَتَفَكَّرُونَ فِى خَلْقِ ٱلسَّمَٰوَٰتِ وَٱلْأَرْضِ رَبَّنَا مَا خَلَقْتَ هَٰذَا بَٰطِلًا سُبْحَٰنَكَ فَقِنَا عَذَابَ ٱلنَّارِ

"There truly are signs in the creation of the heavens and earth, and in the alternation of night and day, for those with understanding (190). Those who honor God in meditation, standing or sitting or lying on their sides, who reflect and contemplate on the creation of the heavens and the earth, (and say) 'Not in vain have You made them. All praise be to You, O Lord, preserve us from the torment of Hell.'"
Quran (3:190, 191)

—

كِتَـٰبٌ أَنزَلْنَـٰهُ إِلَيْكَ مُبَـٰرَكٌ لِّيَدَّبَّرُوٓا۟ ءَايَـٰتِهِۦ وَلِيَتَذَكَّرَ أُو۟لُوا۟ ٱلْأَلْبَـٰبِ

"This is a blessed Scripture which We sent down to you [Muhammad] for people to think about its messages, and for those with understanding to take heed." Quran (38:29)

–

وَٱتَّبِعُوٓا۟ أَحْسَنَ مَآ أُنزِلَ إِلَيْكُم مِّن رَّبِّكُم مِّن قَبْلِ أَن يَأْتِيَكُمُ ٱلْعَذَابُ بَغْتَةً وَأَنتُمْ لَا تَشْعُرُونَ

"And understand and follow the guidance of that which is revealed to you from your Lord, before the doom comes on you suddenly when you do not expect it." Quran (39:55)

SCIENTIFIC EVIDENCE OF PROPHECIES

As stated in the Quran (47:18), when the time for the hour arrives, the command from Allah Ta'ala will descend in faster than a blink of the eye (Quran 54:50) to blow the first trumpet. All who are in the heavens and the earth will stare in fear and will become unconscious save him whom Allah wills (Quran 39:68).

It is a scientific fact that sound waves are as effective as laser rays to physically affect objects. The power of sound can break glass or ignite a fire. Imagine the power of a trumpet blown to the order of Allah. Once it is blown, it will bring the spin of the Earth to a screeching halt, creating sudden jolts, earthquakes, tsunamis, and dust clouds covering the globe (Quran 44:10, 11).

Then the inertia of the Earth's rotation will make it rotate backwards, in the opposite direction. At that time, it will appear as though the sun is rising from the West. Then the second trumpet will be blown and it will be so strong, shattering the already weakened 37-mile-thick crust of Earth, causing it to burst. Mountains will crumble and propel into the air like dust storms, and humans will be scattered like moths (Quran 101:3, 4).

It is now time to appear before The Almighty, The Most Powerful, The Most Beneficent, and The Most Merciful for judgment.

All of this was revealed in the Holy Quran about 1400 years ago, at which time it was unthinkable to even imagine any of these topics. This is also a miracle of the Holy Quran: that the more we study it, the more realities are revealed.

Allah Ta'ala knows that we have much more of the Holy Quran yet to learn. For example:

ٱللَّهُ ٱلَّذِى خَلَقَ سَبْعَ سَمَٰوَٰتٍ وَمِنَ ٱلْأَرْضِ مِثْلَهُنَّ يَتَنَزَّلُ ٱلْأَمْرُ بَيْنَهُنَّ لِتَعْلَمُوٓا۟ أَنَّ ٱللَّهَ عَلَىٰ كُلِّ شَىْءٍ قَدِيرٌ وَأَنَّ ٱللَّهَ قَدْ أَحَاطَ بِكُلِّ شَىْءٍ عِلْمًۢا

"It is Allah who created the seven heavens and of the earth the same number, the Command descending down through all of them, so that you might know that Allah has power over all things and that Allah

encompasses all things in His knowledge." Quran (65:12)

Our search for life elsewhere among other planets in the universes should be over soon, if Allah wishes us to find other life forms. And what else will we discover?

Allah Ta'ala knows the best.

Submitted humbly by:
Ghiasuddin Ahmed Khan
Retired Engineer;
Served, Picker X-Ray (Medical Imaging Systems)
Westinghouse Electric Corporation,
General Motor Corporation in the U.S.A

AUTHOR BIOGRAPHY

Mr. Ghiasuddin Ahmed Khan came from a respectable family from Delhi, India. He earned his Mechanical and Electrical Engineering degrees from India and Glasgow, Scotland. In 1960 he started working as a Chief Engineer in a metal works company. He was in charge of 3 plants, having built one, Zincoxide. He also started his own company, "Technol," in Hyderabad, India. In 1978, he moved to America with his family and worked at Picker International as a senior engineer. After retiring in 1988, he worked at General Motors in Lordstown, Ohio and started working on this book. He passed away in May 2011 before being able to publish it. He is survived by his wife, six kids, twenty-five grandkids, eight great-grand kids and one great-great-granddaughter.

TRIBUTE

Nawab Ghiasuddin Ahmed Khan Sahab, author of this book was the great grandson of Nawab Ziauddaula Bahadur of Delhi who was the prime minister of Mughul King Akbar Shah II. Nawab Ghiasuddin Ahmed Khan, besides being a respected professional engineer, was a thinker and a God-fearing man who loved to investigate Qur'anic verses to discover their true meanings and relationship with human lives. He firmly believed that the purpose of the Qur'an is to mold human character and life style for success in this life and the Hereafter.

Jamaluddin Ahmed Khan,
His youngest brother,
Retired Staff Sergeant
Toronto Police Service,
Canada.

NOTES

NIYAH PRESS

Do you have a story to tell?
*Join us at **NiyahPress.com** to learn more about*
how we can help you share your message with
the world.

OTHER NIYAH TITLES

The Spiritual Adam by Imam Abdullah El-Amin
The Black Mzungu by Dr. Alexandria Osborne
Jihad of the Soul by Zarinah El-Amin Naeem

Like Glue: The Little Book of Marriage Advice we should have stuck to from the beginning by Dr. Halim Naeem and Zarinah El-Amin Naeem

A Part of Me Refused to Die by Nisha Sulthana
Muslim in Transit by Mohammed Qamruzzaman

Follow us on Facebook for publishing tips.

www.niyahpress.com
info@niyahpress.com

Made in the USA
Monee, IL
20 May 2022